THE SECRET OF LIFE

ROBERT DOMIENIK

authorHOUSE®

AuthorHouse™ UK Ltd.
500 Avebury Boulevard
Central Milton Keynes, MK9 2BE
www.authorhouse.co.uk
Phone: 08001974150

First published by AuthorHouse 5/19/2009

ISBN: 978-1-4389-7531-3 (sc)

This book is printed on acid-free paper.

Contents

INTRODUCTION

When I was looking for a way to sort out some processes connected with the evolution of the solar system for example, process the   process of creation of moons or comets, I met a problem which is that it is impossible because of the rules of dynamics. The rules of dynamics are certainly correct, so what is wrong? Let's sort out the information based on scientific facts and not researched prediction. As an example I will use our planet earth. The natural weight of our planet is 5.5kg/litre. The natural weight of rocks is 2.5kg/litre, the natural weight of iron is nearly 8kg/litre and they cannot exist in any other form. In a liquid state mixing is impossible. If we ignore scientific fact, liquid iron is as good a conductor of mechanical wave as liquid rocks. We could make our planet from liquid rocks with an iron core or we didn't ignore this fact and make our planet from liquid iron with floating rocks on the top and gas core in the middle, I choose the second option, I would like to remind you here of the often-forgotten scientific fact that an astronomical body is not a source of gravitational force but only each separate particle is a source of gravitational force. Let's analyze the potential through a theoretical body which contains only two elements of hydrogen. The basic condition here is that there exists a distance between them all the time and in all circumstances. Let's agree that the direction of the potential of the field directed towards the centre of the body is called "positive" while that directed outward is called "negative". It will be noted here that the potential of the field between the centre and nearest particle is always negative. We can also set a border between a negative and positive gravity force field. In my opinion the increase of the number of particles in our body is the reason for expansion of the internal zone (negative gravitational force). Research on our moon (compare theoretical accounted capacity to real capacity) confirms these phenomena. If we agree with the fact that; negative gravitational force is responsible for empty space in every astronomical body (excluding iron magnetic bodies) we have got one problem already sorted out. I started to think about how to sort out the secret of life from this idea: Energy from chemical bonds of the compound water may by enough to destroy a planet. I wonder where this energy is. The most interesting and secret is the chemical compound hydrocarbon. Where and how did it arise? The next problem for my attention is the statistical composition of the elements in various places. Why is it so similar in moon rock in earth rock in DNA? How I deal with these problems I'm going to leave for my reader's judgment?

Add. Illustrations

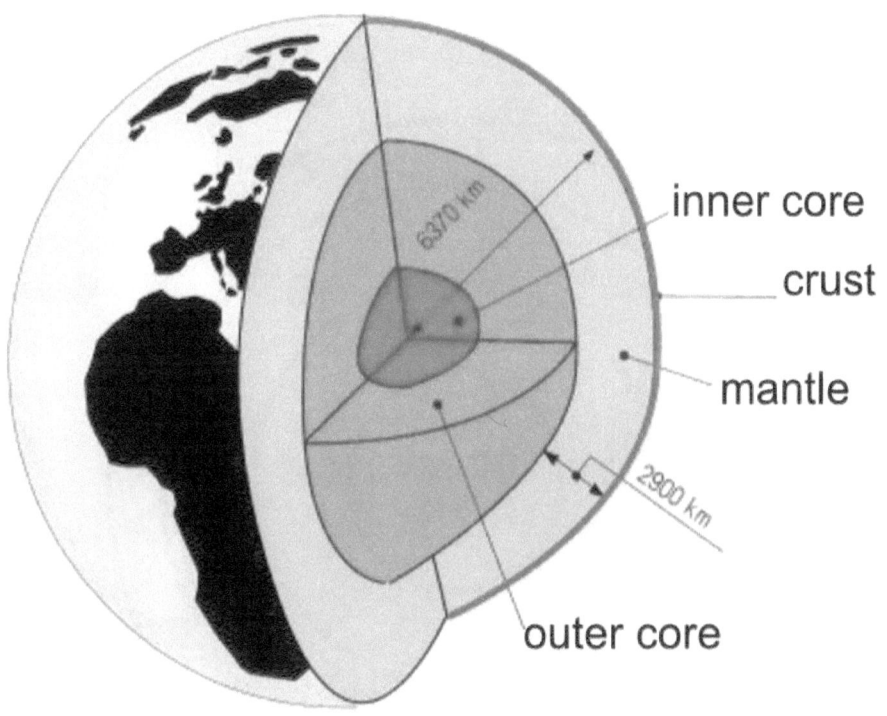

For human civilization it is easier to land on Mars then to rich 50 km (less then 1% of radius) deep in our planet Earth.

Fokus of earthquake

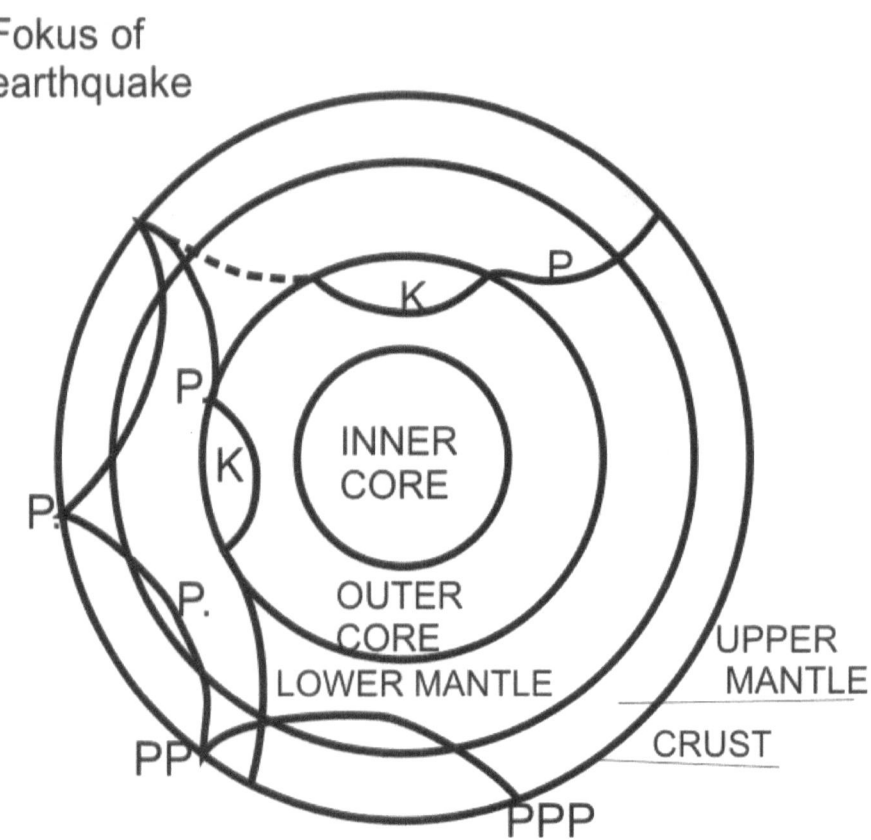

Seismological research shows us that core of our planet does not conduct mechanical waves. Only gas or a vacuum is not able to transport energy by the mechanical waves.

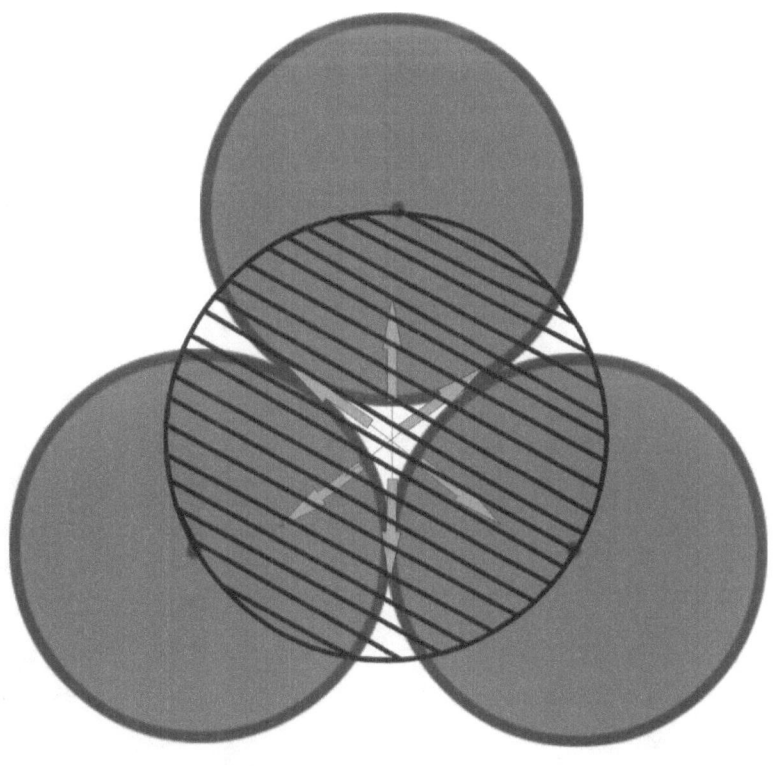

 - Magnetic force field
ensure distains between particles

- Inner zone with gravitational
force directed outward the centre

Magnetic force field of particles ensures distance between them
ALWAYS and EVERYWHERE

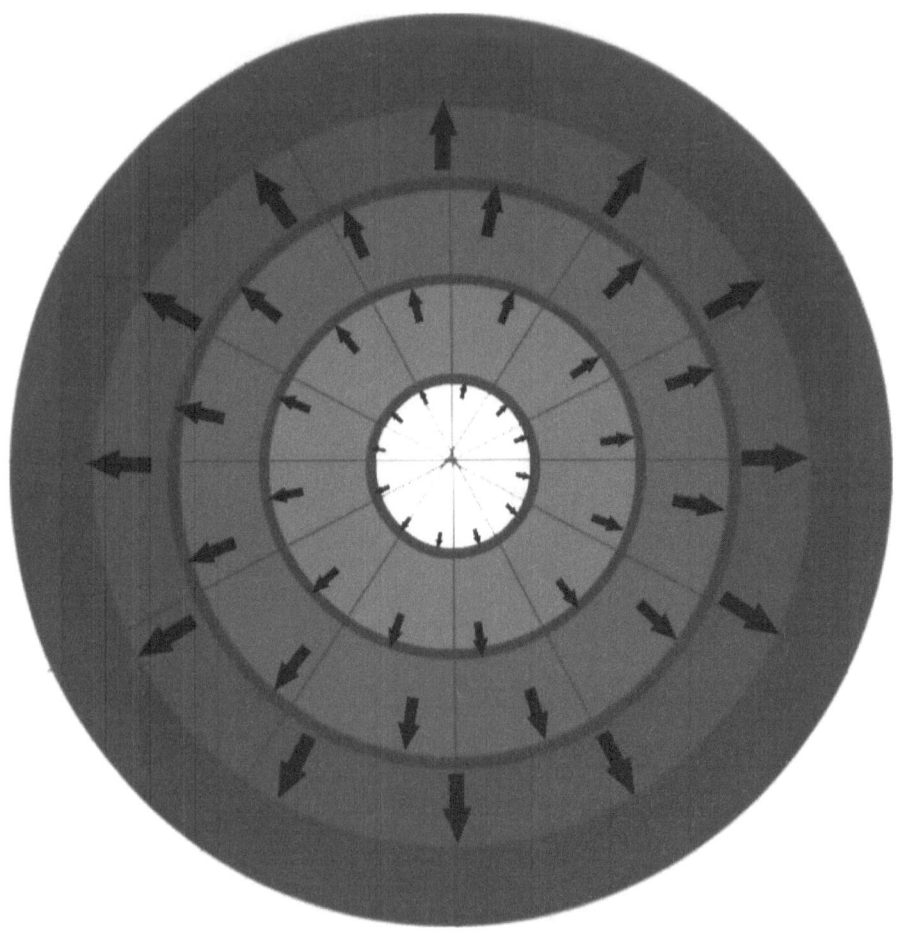

Gravitational force directed outwards from the centre acting always whenever a body has size.

Elements

The fact that the elements exist in the solar system ought to be enough to prove that they arose from the sun. If their speed was fast enough high to leave gravitational force of a star after an explosion like a supernova it should stay high enough to leave a gravitational force of our star.

We haven't got any atmosphere in the solar system to slow down the speed. From research about the chemical contain of our son we know that there are 1000 times more elements of iron then weight of our planet earth. Known distances between the stars mean it is absolutely impossible to get such number of elements from them.

Now I will consider the causes and the course of a supernova star explosion. The explanation that the end of the element hydrogen is the reason for such explosion I do not accept because it is not leading from reason to results. Only an increase a certain level of energy can be a reason for such an explosion. Bearing in mind the size of our sun and remembering about the inside zone with negative gravitational force our sun is looks like a sort of bulb covered with a coat of a certain thickness. The nuclear process takes place in a thin zone containing the element of hydrogen

Where this occurs in solid state, the solid state of this element is a necessary condition for a nuclear process (chain reaction) to occur. The existence of a solid state of the element hydrogen is limited by temperature from the side of high. Increase temperature changing state from solid to liquid it is the reason to fall down the chain reaction. A reason to start the nuclear reaction can be a superposition of the mechanical wave. In my subjective opinion the reason for an explosion like a supernova can be extraordinary event like very big nuclear explosion. The event in consequence is similar as if one brick was taken away from the arch in a vault. The balance of forces is disturbed and the vault collapses. I picture this process as follows: A funnel forms in the low pressure zone arising as result of the nuclear explosion; this event looks like the star has poured inside itself through this funnel. Have in mind size and dimensions of the star. This process takes quiet a long time. Simultaneously, the increase in the potential of the gravitational force acting along the diameter depresses the arch of the star on its opposite side. A new funnel is then formed through which the star pours inside itself. Both funnels increase in size and ever greater areas of the star collapse. The speed and inertia of masses speeding against each other grow enormously. They collide. Energy released in this process is sufficient for the process of synthesis of silicon into iron to take place.

I would like to throw some light on this process. As shown by research done in accelerators, apart from the ignition energy (the energy necessary to trigger the process) a nuclear reaction possesses overheating energy. This energy is so huge that it makes it impossible for the process of nuclear synthesis to take place. In particular, in these conditions it is possible for a proton to pass through the nucleus without being trapped by nuclear forces. What does it mean in practice? It means that only a part of matter can take part in nuclear synthesis processes. And this refers only to oxygen and silicon group elements; the elements helium and hydrogen will not take part in the process of synthesis at all for these conditions are highly unfavorable for any synthesis of these elements to take place. The products of this reaction will be elements we are well familiar with as well as a substantial number of their isotopes characterized by only a slightly altered statistical composition Another consequence of the explosion is, obviously, speed. An iron particle has no definite speed just after the reaction, its motion resembles shaking, arising from the instability of the inside of the nucleus. It is only by colliding with another particle that it gains speed. The particle than transmits the speed to another particle and then to yet another one. That is how the mechanical shock wave originates. The wave has the highest speed produced during the reaction. This way only part of the matter is thrown away into space. As is the case with any other explosion, the speed at which most of matter is thrown away into space is a result of the increase in pressure around the place of reaction. The amount of energy produced during the reaction is huge, but it is to be remembered that the amount of matter imparted with that energy is huge, too. While determining the speed at which matter is thrown away into space, it is worth discussing the subject with people conducting research on nuclear weapons. I think we are talking here about the speed of several hundred kilometers per second. And this speed is definitely not sufficient for any matter to leave the gravitational field of its mother star. A halo moving away from a supernova star is certainly not any form of matter. No material thrown away after an explosion emits light for there is no reason for this to happen (there are no processes taking place inside the matter that could be the source of light). What we observe is the emission of light induced by gamma waves acting on particles that are already present in the vacuum. This phenomenon clearly shows that interstellar vacuum is not an absolute vacuum. At this point I come to the conclusion that stars explode many times as well as to another bold conclusion, to the effect that every stable star has a planetary system. I believe that the presence of heavy elements in a star is the condition of its stability. And so we have almost the same matter as before, but now it is dispersed over a large area. What will it become? There are several options: a star with a planetary system, a double star system, a pulsar star, a system of pulsars and a dead star. A dead star is a formation similar to our planet Jupiter, though it can be much bigger. However, its size is not sufficient for nuclear processes to be triggered. It forms from helium and hydrogen that do not take part in the reaction in explosion as supernova. While it does not emit light, it can block the light of another star, much to astronomers' joy.

Let us have a closer look at the option "a star with a planetary system". The most interesting question to me is this: "When and how did the chemical compounds originate?" This question is of key importance to energy balance in the process of planetary formation. What is particularly interesting here is planet Earth and planet Venus. Let us start from the explosion itself. Are the conditions during the explosion suitable for chemical compounds to originate? Certainly not. The next event to take place is particles' moving away from the epicenter. They travel along virtually parallel trajectories moving away from each other. The temperature of all the matter falls to absolute zero.

These conditions are not favorable either for any chemical compounds to originate. Let us see what happens to the elements.

In my reasoning I follow Isaac Newton's laws of motion, which I regard as the only true and irrefutable laws of motion. I accept a preliminary assumption that the energy of the reaction of nuclear synthesis of silicon into iron was transmitted to all matter as a result of perfectly elastic collisions between molecules. Newton's laws of motion tell us that the energy level of molecules does not change as they move away from the epicenter. As it is known, the speed of different molecules is not the same. Rather, it depends on their individual atomic mass. Let us try to locate five deliberately selected elements using the statistical distribution of speed according to Maxwel-Boltzman's speed distribution for gases. The elements whose position I am interested in are: iron (with the value of 56), silicon (with the value of 28), oxygen (with the value of 16), helium (with the value of 4) and hydrogen (with the value of 1).

The element to be found closest to the Sun is iron, being the heaviest element with the lowest initial speed. The next in line are silicon, oxygen, helium and hydrogen. The planet Mars indicates a place where matter's density is relatively low because of the planet's small size.

I believe this is the place where iron's range of occurrence ends and silicon's range of occurrence begins. If planets Venus and Earth originated from a single group of elements, their identical sizes are a consequence of the symmetrical distribution of speed according to the chaos theory. To be precise, the process of iron consolidation begins from the effect of the gravitational field of the first and biggest planet in the system, Jupiter, and the Sun. Internal gravitational forces of iron begin to act just after the preliminary formation of a very oblate ring. When viewed in cross-section, this ring is much elongated as well as symmetrical, in accordance with the theory. That is why its own gravitational field tears it into two identical parts, similar to a thin trickle of water being torn into drops by the magnetic force of a water particle. Let us now move beyond planet Mars.

This is the right time to say something about the quantitative statistics of elements sought. In terms of quantitative statistics I can say that on the Sun that there is a similar amount of iron and silicon, about two times more oxygen than silicon while such elements as helium and hydrogen are in a class by themselves, their amount being many times larger than that of heavy elements. Coming back to chaos theory, if silicon elements begin to occur in the area of Mars, their highest concentration should be found in the asteroid belt. Oxygen could begin to occur in the same area as silicon, but its highest concentration will be found closer to planet Jupiter. As for helium and hydrogen, the highest concentration of helium is in the area of Jupiter, while that of hydrogen is shifted towards planet Saturn. Given the fact that they occur in much higher amounts, their area of occurrence might begin in the same place as that of oxygen and silicon.

As the above analysis shows, there is an oxygen-silicon planet missing in our solar system. Let me here remind the reader that the elements of the oxygen group are: carbon (with the value of 12), nitrogen (with the value of 14), oxygen (with the value of 16); and elements of the silicon group are: magnesium (with the value of 24), soda (with the value of 23), aluminum (with the value of 27), phosphate (with the value of 31), sulphure (with the value of 32), silicon (with the value of 28), chlorine (with the value of 35), potassium (with the value of 39), calcium (with the value of 40).

speed of entering

acceleration

sun

slowing down

speed of leaving

Speed of entry is the same as speed of departure.

Because of this all the matter in the solar system can be originate from the sun only.

Sun operates like a diesel engine, one explosion is the reason for the other.

Focus of the nuclear chain reaction transforms into a low pressure zone, where temperatures slightly drop down and it is visible as a dark drop on the surface of the sun.

This is picture of the explosion on the sun. It can be the effect of a nuclear chain reaction only.

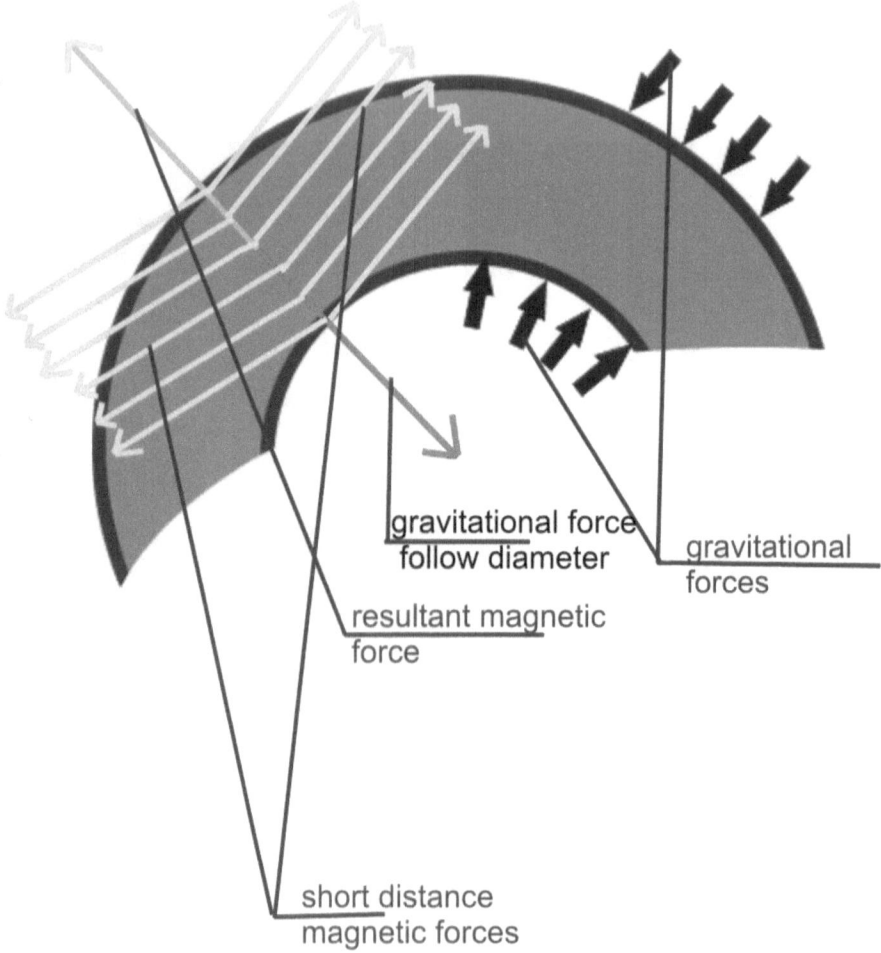

gravitational force
follow diameter

gravitational
forces

resultant magnetic
force

short distance
magnetic forces

The balance of the forces on the sun is similar to the balance of forces in the bridge

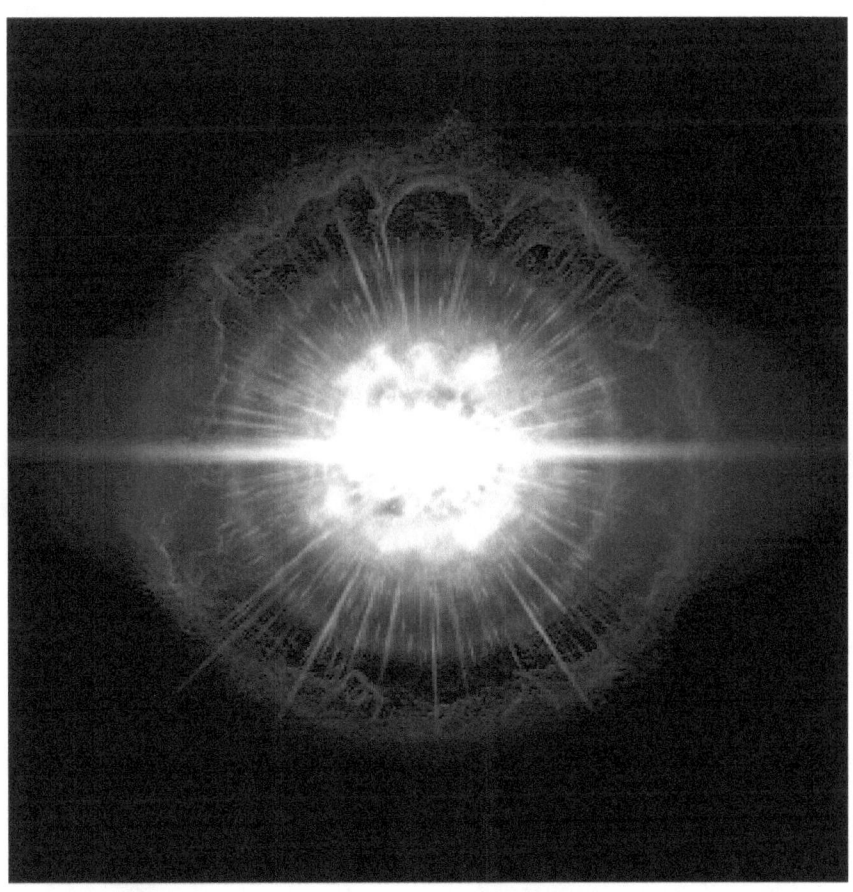

Explosion like a supernova create speed smaller then human in laboratory called accelerator.

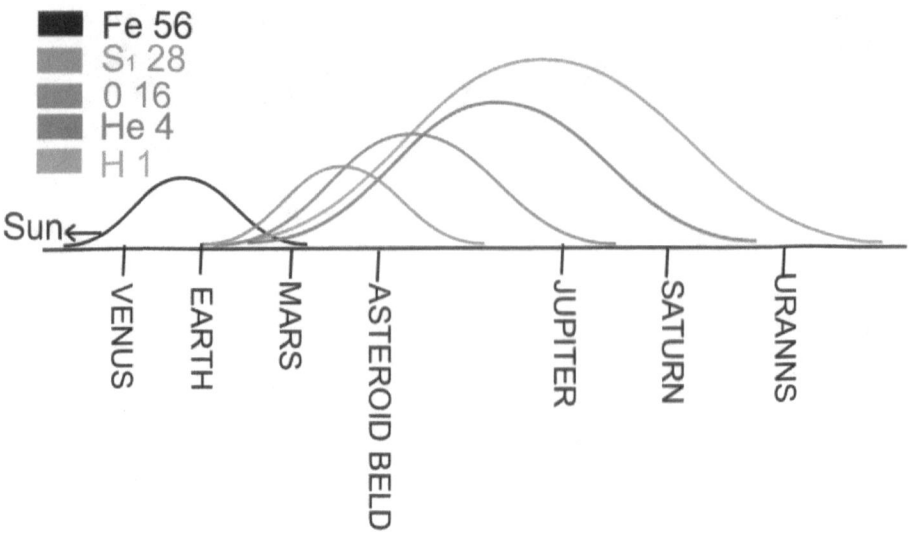

Distribution of the elements in space is in accordance with Maxwell-
Boltzman theory about speed distribution.

A non-existent planet

I created a planet out of chaos theory. Its atmosphere was composed of hydrogen and carbon, a sea of liquid nitrogen stretched underneath. The bottom of the sea was made of oxygen in a solid state. The closer to the centre of the planet, the more particles of aluminum, sodium, manganese, phosphate, magnesium, silicon, potassium and calcium would be found. Our theoretical solar system is perfectly quiet. There are no meteorites, no asteroids, and no comets – there is nothing to disturb the peace. Nothing comes to my mind. How did it start? I put that problem aside to deal with it later. And then one day I watch an old TV documentary series presenting the achievements of the Mariner and Voyager missions. There is the information that a volcano was discovered on the moon IO in the Jupiter system. The explanation of this phenomenon is as follows. Each of the three great moons of Jupiter is characterized by the different length of time it takes to orbit around its mother planet. Consequently, every imaginable position of one of them in relation to all others is not only possible, but certain. It is only a matter of time. It happens on a regular basis that all of them are arranged in one line. In those occasions, the potential of an external field acting on the inner moon and trying to deform it is slightly different from usual.

This situation immediately brought to mind my oxygen-silicon planet (as I call it). The planet could easily form because planets Saturn, Neptune and Uranus came into being as the latest ones. After their formation, it is certainly only a matter of time before the last five planets find themselves positioned in one line. In my opinion, carbon-hydrogen compounds are the first compounds to have formed in our solar system. So did it all start with a storm in the carbon- hydrogen atmosphere? It could not have been a violent eruption. Because of a high content of helium, it was more likely a slow process of burning out. The energy that has originated in the process brings nitrogen to the boil.

Subsequently, it penetrates ever deeper layers of our planet. The reaction of synthesis with the silicon group elements is so violent that the outer layer of the planet is thrown away into space. A shock wave generated by the process triggers the reaction of nuclear synthesis of silicon into iron. The wave hits the empty inside of the planet leading to an enormous increase in the internal pressure. The pressure causes the planet to burst. Subsequently, cracks form through which gases escape at unimaginably high speeds. In my opinion this is the moment when comets originate. All that remains is a small incandescent part of the planet heated by the nuclear synthesis reaction to a temperature of over ten thousand degrees. According to the laws of dynamics, the change in the position of matter in relation to the axis of rotation leads to a significant increase in its rotational speed. Gravitational force, much weaker now than it was initially, is not able to counteract centrifugal force. The agony of the planet looks like this: its incandescent remains are thrown around in the form of huge gas bubbles. Some of them are thrown forward in relation to the original direction of the planet in its motion around the sun and some are thrown backwards. This is the rule of dynamic. As a consequence, those thrown forward gain speed in relation to the sun and those thrown backward slow down in their motion around the sun. I suggest we have a closer look at ten of them. The five that were thrown forward move along a spiral trajectory in the direction of the more distant areas of the solar system. This movement is caused, obviously, by an increase in centrifugal force connected with speed. As they move away, their speed falls and the spiral trajectory becomes denser. Four of them are trapped by the gravitational field of planet Jupiter. These are Io, Europe, Ganymede and Calypso. The fifth is trapped by Saturn's gravitational field. This is Titan. Let us now look for their counterparts in the part of our solar system closer to the Sun.

One of them is Mercury, wrongly called a planet. The other one is the moon orbiting around our planet. And where are the three remaining moons? The only reminder of these is the energy they have left behind. I suggest you calculate the energy of difference of the highest and lowest trajectory of planet Venus or planet Earth and then compare this energy with the kinetic energy of the difference in speed between a middle-sized moon and planet.

Add. Illustrations

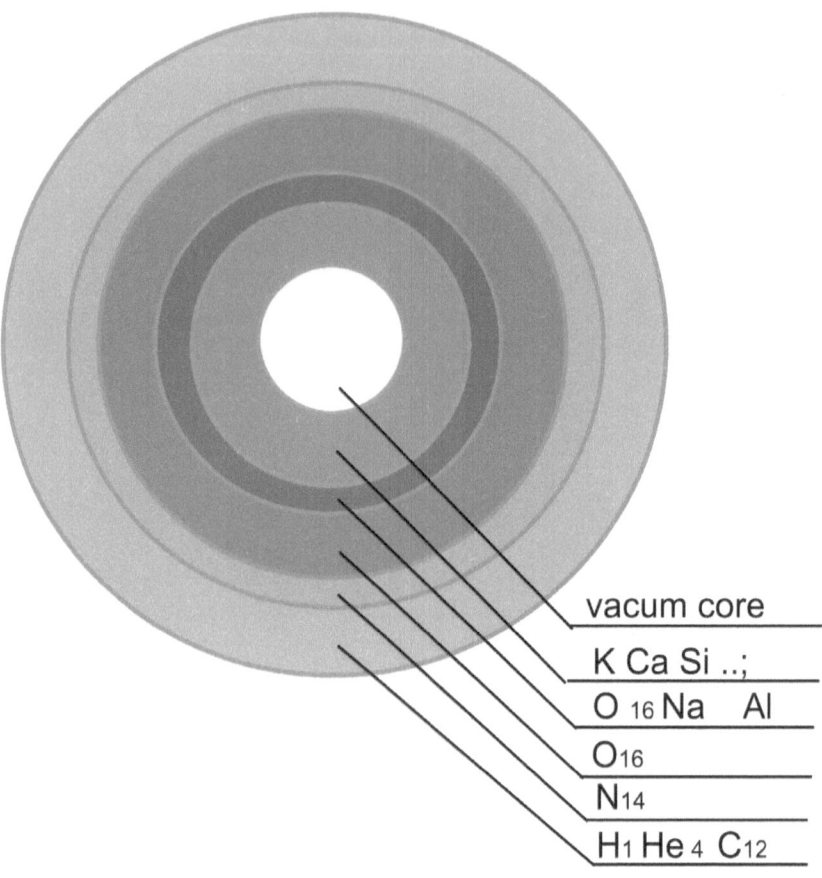

vacum core

K Ca Si ..;

O $_{16}$ Na   Al

O$_{16}$

N$_{14}$

H$_1$ He $_4$ C$_{12}$

Among the elements: Hydrogen, Carbon and Oxygen, the chemical connection Carbon-hydrogen is the weakest one. The value of the atomic mass sets those elements closer to themselves than to the Oxygen only.

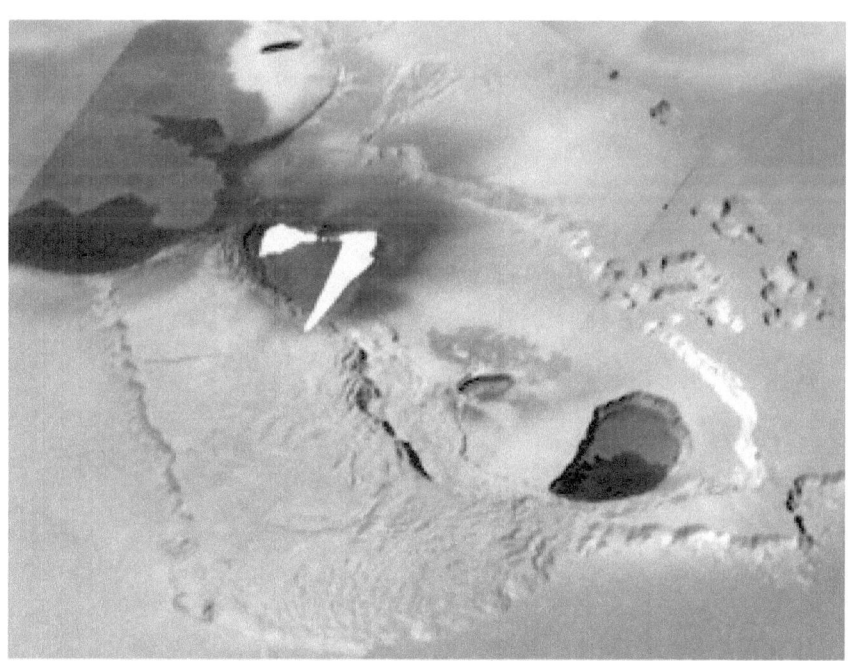

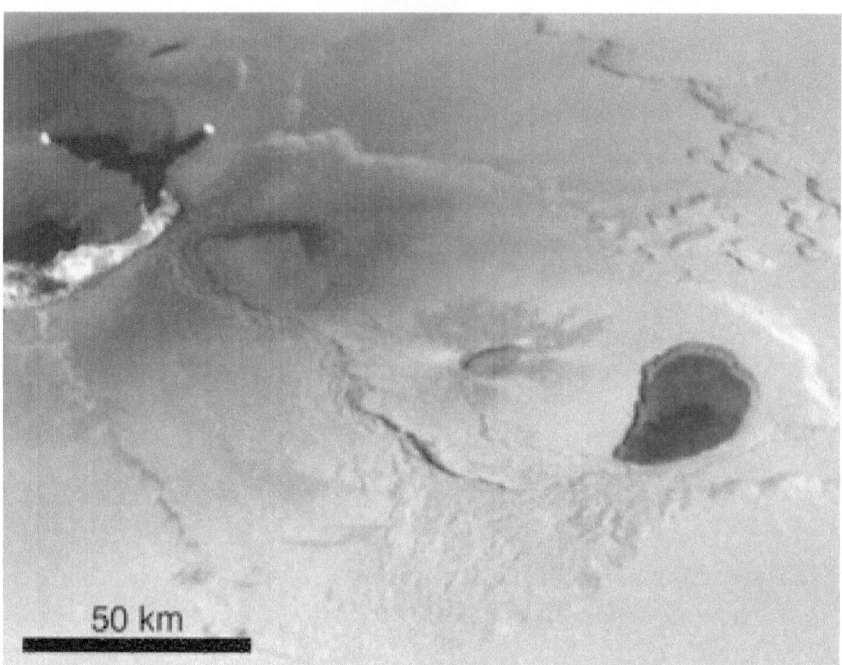

Volcano on the moon Io, pictured by NASA in mission of Voyager 2

This is Comet. Its chemical composition tells us that it is originated from a now non existent planet. It speed tells us that it is originated from explosion.

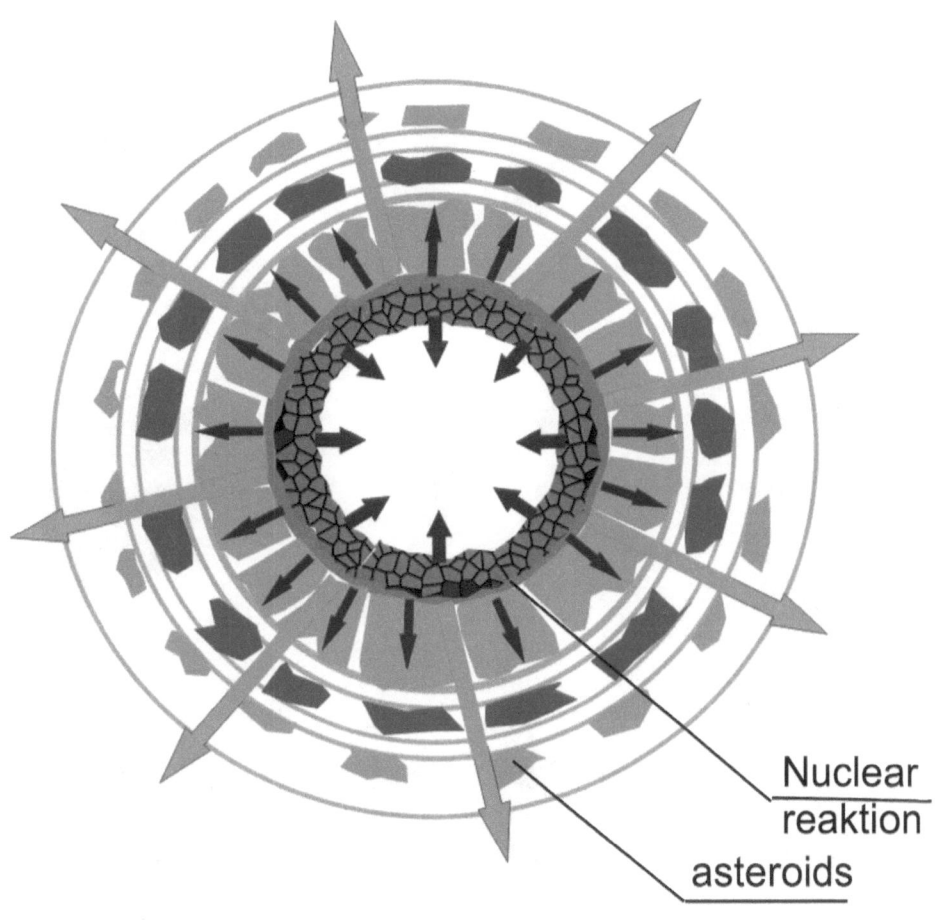

Nuclear
reaktion
asteroids

Increase pressure in empty core of the planet to blow it up from inside.

The fact that an explosion of the planet took place confirms phenomena that asteroids and meteorites similar in size move with the same speed.

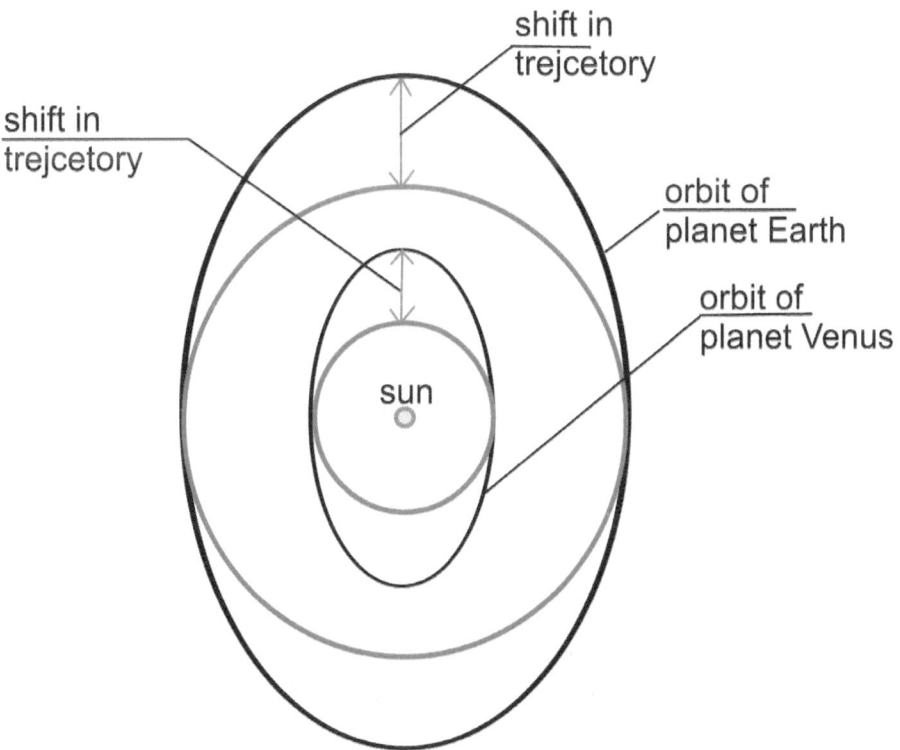

Shift in trajectory of the planets Earth and Venus is the effect of the collisions with the moons.

Planet Earth

Follow the theory of chaos our planet is a solid, black, cold and perfectly magnetic bowl. Without atmospheric water, nothing disturbs its perfect shape. Four and half billion years ago an amazing event took place. The night sky was lit up by five suns, the same size and same brightness as our sun. They moved more slowly than our planet so our planet caught up with them. One was trapped by the gravitational force of our planet one collided and three disappeared toward the sun. As a result of this collision there arose an ocean on our planet. It was a thick, viscous liquid similar in consistency to our blood. The temperature of the ocean was also the same, that is, about 40 degrees. It was most probably white in color, due to the high content of calcium and magnesium compounds. Each drop of this ocean was different; each contained billions of newly formed biological structures. During the day, under the influence of solar radiation, these structures partly disintegrated to create new, stronger configurations at night, when the temperature fell slightly. Apart from the ocean, there also was an oval continent on the planet. It was regularly elliptical in shape and situated in such a way that its longer diameter ran parallel to the planet's parallels of latitude. Its elongation is a consequence of our planet's rotation around its axis. Before long, the regular oval shape of the continent transforms. Energy from just-created polymers warms up our iron planet from minus 200 plus to 40 degrees centigrade (the structure of DNA is a polymer. The chemical compound DNA is on the one side explosive like dynamite and on the other side it is very delicate can - it be irreversibly destroyed in temperatures below 50 degrees centigrade). Chemical reactions between oxides, nitrates, carbonates and silicon group elements (which resulted in the creation of today's rocks) produced also thermal energy. As this energy moved deeper inside our planet, it tore the magnetic bonds between iron particles and changed iron's state of aggregation from solid to liquid. The whole force system of the planet changed. As a consequence of this process, the force of the magnetic field, until then many times stronger than the gravitational force, was no longer dominant. It was the latter force that began to determine the shape of our planet. Its effects are completely different. In consequence of the above, our planet expands. Upward currents of liquid iron fill the growing area of our planet and tear apart the continental block, which is how the Atlantic Ocean comes to be. The shape of Canada's eastern coast is also a result of this process, as are the Cordeliers, the only mountain range on our planet to have originated from tectonic block movements. Varied in its topography, our planet was shaped as a result of a collision with asteroids. I will try to present this process using one of them as an example. (I will take this opportunity to add that the chemical composition of an asteroid and a moon is the same). It happened in the pre-Cretaceous period, if we go by the geological calendar. An enormous asteroid with the diameter of several hundred kilometers was approaching the Euro-Asia continental block from east toward its north-western part following a track parallel to the planet's parallels of latitude. The angle at which it hit the planet was similar to that of a landing plane. Its passage through the atmosphere was very short. Let us bear in mind that the asteroid is perfectly cold and that it is mainly composed of silicon oxides. It is necessary to apply huge quantities of energy to decompose this compound and so the asteroid is very damage-resistant. It touches the land in the place of today's Finnish Bay. As a result of this impact, a continental block breaks up giving rise to the Ural Mountains. The asteroid then continues to tear through the block giving it the shape of a huge furrow. The northern edge of this furrow becomes the Scandinavian Mountains while the southern, the Carpathian Mountains. Just ahead of what today is the Danish Peninsula, the asteroid drowns in the depths of liquid iron. And because it is made of material that is three times lighter than iron, before long it resurfaces to become an island which today is called Great Britain.

To conclude this part I would like to start a scientific investigation on the subject of the greatest space catastrophe in the history of our planet.

The primordial rocks of the Earth's crust, rich in silicon oxides and light in color are acid, called granites. In my opinion, they are of lunar origin, whereas the black rocks, called alkaline, constitute the primordial crust of our planet from the time before it came in contact with the moon. Consequently, light rocks are always to be found close to the surface of the continental block while black alkaline rocks are to be found in the deepest layer of the continental block as well as under the ocean bed. I was intrigued to read in a geological publication the information that on the eastern slopes of the mountains in eastern Australia this order is reversed, that is, black alkaline rocks are to be found on top and light acidic rocks beneath them. We know practically all the mysterious forces governing our planet. Which of these forces, then, is responsible for turning a tectonic block upside down on such a spectacular scale? Could it be the same force that uplifted the Himalayan range and superimposed one block on top of another to create the Tibetan Upland? I leave this investigation to be continued by the readers.

Add Illustration

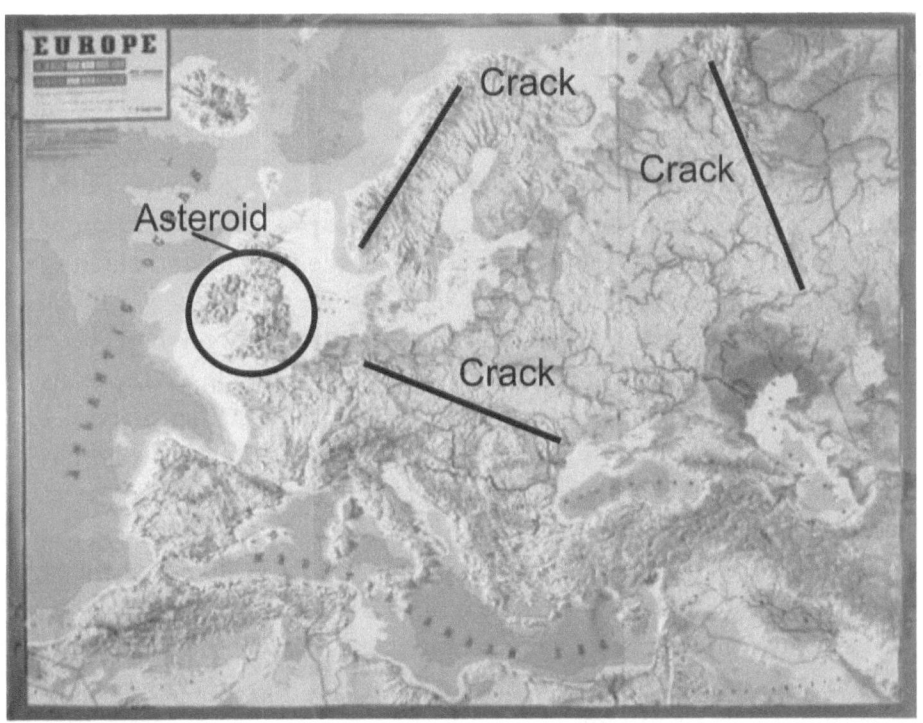

Conclusion

I wonder if it is possible to find today and, indeed, if it exists, an irrefutable proof that the solar system once witnessed a reaction of nuclear synthesis outside the Sun. Opening the Internet and entering "chemical compound DNA structure". I can see the model of a DNA particle. I can have no doubts: that this is the proof I have been seeking. Looking at this picture I imagine atoms racing in all directions and suddenly something very strange happen: time suddenly stops. I asked myself the following question: What physical process or what phenomenon is capable of stopping the time? There is no such process and no such phenomenon. But wait a minute - what is time? It is a measure of an event as it happens. In this case it is the movement of particles that plays the role of such an event. And so the question is: what physical process or phenomenon is capable of excluding speed? I think I know. It is a collision, a collision with something huge and motionless. If I had money and a laboratory I would try to produce such a particle by myself. And this is how I would go about it. I would prepare a huge iron bar cast in such a way that it would set under the influence of a very strong magnetic field. In this way I would obtain the best possible ordering of atoms in terms of their magnetic orientation. The bar should also have cooling channels with liquid helium running inside it. I would place this rod in a big vacuum chamber and I would connect it to a cooling system outside the bar. The chamber would be equipped with a nozzle resistant to very high temperatures and with its end directed towards the bar. After cooling the bar to the lowest possible temperature, around minus 200 degrees, I suppose, I would use the nozzle to introduce a gas with a statistical chemical composition similar to existing biological structures. It should be remembered that it is just DNA we are examining, not an organism (it is necessary to dehydrate the organism). The temperature of the DNA should be high enough to eliminate the possibility of any chemical compounds coming into existence; I think 3000 degrees Centigrade should be enough. At the end of the experiment I would fill the vacuum chamber with steam. And then I would examine the outcome of the experiment. I owe you an explanation as to where the water is from. One source is waste product of biological processes; second source is importing of element hydrogen from our moon. Also I want to remind you here that our moon contains 43% of oxygen. On the surface of our planet this element occurs as often. Even now I have an idea how to use the new DNA particle. There was a time in the history of our planet, in the Carbon Era to be exact, when plants growing on Earth eliminated carbon dioxide from Earth's biosphere. This led to the cooling of the climate and brought about the destruction of these plants as well as the whole ecosystem they were part of. Today we have to find, among existing biological or artificially made structures the ones that will be able effectively to purify Earth's biosphere from carbon dioxide. This is the necessary condition for the survival of biology on our planet. No Earth ecosystem can play this role for it came to be (was selected) in the chemical conditions that existed on Earth before human civilization altered them (what I mean here is the massive emission of carbon dioxide). Existing Earth ecosystems are not suited to changing conditions and they will not be able to adjust to them. Biological structures do not come to be; rather, they are selected. They have only come into existence once on our planet. Diversity is biology's strong point. It is diversity that ensures biological and chemical balance on our planet. It seems to me that the only way to restore the chemical balance in Earth's biosphere is to enrich our planet with new biological structures. In my opinion it is the cheapest, the simplest and, I'm afraid, the only way for life survive on our planet, Earth.

Add. Illustrations

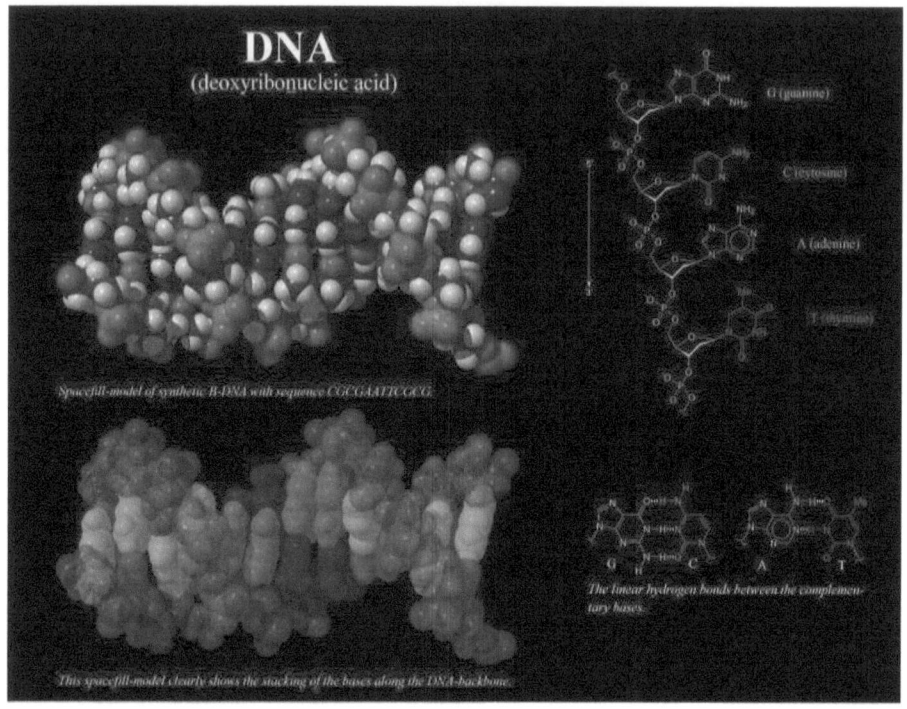

DNA can not arise from chemical compounds only from separate elements.

DNA was selected from an infinity number of different chemical compounds as the best and strongest.

SUPPLEMENT

I would like to say something here about phenomena which confirm my theory. One of these phenomena is pictured by NASA the Kuiper Belt. From my description about the explosion of our sun as a supernova we know about two different speeds arising during this process. The faster one is from a mechanical wave which in this way was throwing out some particles of matter which occur today beyond the planets called the Kuiper Belt.

In Kuiper Belt we can find also a matter from explosion of the planet, .Another movement arose because of the increase in pressure around the nuclear reaction and the matter thrown out in this way occurs today as planets. If we agree that planets Earth and Venus are made of iron we get the distribution of elements perfectly according to the speed distribution follow Maxwell & Boltzmann theory. My description about how was arise chemical compound hydro-carbon is confirmed by discover it on the moon Titan in Saturn system.

Now I would like to explain why I do not accept that the thermal energy in our planet is from collapsing process.

Explanation is required so we picture a solar system the scale of which is suited to human perception. My way of doing this is that I shorten all sizes and distances in the solar system by 100.000.000. Now the solar system looks like this: our sun is a bowl with diameter 1.4 metres. The planet Venus is the size like cherry, about 1 centimetre and its distance from the sun is about 1 kilometre. Planet Earth is a similar size and its distance from the sun is about 1.5 kilometres. Jupiter is like an orange 7.7 centimetres across and its distance from the sun is about 8 kilometres. Planet Uranus is 5.5 centimetres across and it is

36

about 30 kilometres from the sun. Now we can better imagine how small planets are in comparison to the distances between them. In this picture it seems our planets were formed in a nearly perfect vacuum. Thermal energy couldn't have arisen in the formulation process of the planets, because movement of particles was so ordered that all energy from movement towards the sun was changed into speed around the sun, which today for our planet is about 30 kilometres per second. All energy from movement in direction perpendicular to the solar system's plane was changed into circulation around its axis. The process of forming of the planets was very slow. I would like to remind here that every collision between particles is the reason of the light emission. So any thermal energy arising in this process had plenty of time to be emitted as light. I would like to say something about where the idea about an explosion in the solar system is from. Look at the picture of the surface of the planet Mars or the surface of the moon, and try to explain that what you see any other way. Also the speed of asteroids, meteorites and comets has to have some explanation. I think we can't explain these phenomena with respect to the laws of dynamics any other way.

Now I would like to draw attention to the shape of our moon and shape of Earth; how is they are different? Our moon is formed of matter in solid state and asteroids after their collisions leave marks like craters but planet Earth is from liquid iron and asteroids after collision leave marks like craters and also leave shapes like volcanoes or mountains after they resurfaced.

One day my son asked me to explain how it happened that we have a volcanic mountain like Kilimanjaro. I told him 'I'll show you' and I took him to the bathroom, filled up the bath with water and put newspaper on the water. Next I took a small ping-pong ball and I sink it down under the newspaper. Next I was let go resurface.

In front of our eyes arose a beautiful volcano shape. My son said it was amazing. You can do simple research like that and after ask yourself or anybody, does it possible to do that other way?

There was slight earthquake in UK last year. I would like to take this opportunity to share my knowledge on this and any other earthquake. Explanation that: The tension in tectonic board is reason of earthquake I do not accept because tectonic board is not made of steel only of stone. Stone is absolutely not able accumulate any tension. Most earthquakes are connected with the internal structure of the asteroid. Before I discuss this subject I would like to make you familiar with the course of asteroid hitting a planet.

This event is similar to what happens when a pane of glass is shot through with a bullet. The asteroid moves at such an enormous speed that it is able to go through the tectonic block and land in the liquid iron zone of our planet. It then resurface under the block and pushes up a cone forming a mountain or a volcano on its surface. I will now move on to discuss the structure of the asteroid. Its chemical composition is identical is with the composition of the moon. The only difference is that on the moon chemical substances are very thoroughly mixed with each other while in the asteroid they from separate groups. Such compounds as carbon dioxides and sulphur dioxide are often found in groups. When the asteroid heats up, these compounds turn into gas and from gas chambers inside it. When viewed in cross-section. The asteroid looks like hard cheese with holes. From underneath the asteroid is washed out by a very quiet and slow current of liquid iron flowing east to west. It originates as a result of our plant's rotating around its axis. The effect of this current is rock material thrown to the surface on the west side. This asteroid is one the oldest in the world. Its age is proved by the sediments of the Baltic Sea basin that the asteroid created. All the gas chambers of asteroid have long been filled with liquid iron. The cause of the earthquake in Britain last year was "collapsing" of a ceiling or a wall of such a chamber in the upward direction. This is the slightest type of earthquake that there is and the island is without a doubt not at a risk of any other type of earthquake. A more serious risk is posed by asteroids whose gas chambers are not filled with liquid iron but sulphur and carbon dioxides. As a result of bottom-up erosion process, iron gets into these chambers. Under the pressure of liquid iron, the gases that the chamber contains escape to the adjoining chamber and then to the next one and so on. Each time they escape their pressure increases. The pressure can increase to such a level that the asteroid together with the mountain and volcano above it might erupt. This is what happened in British Columbia in Canada – I mean mount Saint Helena. Most often, however, these gases escape in a more or less stormy manner through the tip of the volcano. To fully cover the subject of earthquakes I will present their last cause. These are the earthquakes with the most dangerous consequences. They are connected with faults in tectonic blocks, movement of these blocks in relation to each other and occurrence in several places at the same time, for example in California and Japan (the famous earthquake in Kioto in Japan). The causes of these earthquakes are gravitational tides, the same that are responsible for the changes in the ocean water level. It is to be remembered that out planet has liquid consistency and also that it is empty inside. This makes it very sensitive to any changes in the external gravitational field.

Add. Illustrations

Pluto and Kiuper-Belt are originated from the explosion of non-existent planet.

Moon like Earth was underwent same collisions but their topography is different.

Natural force directed from down to up is most often Force of displacement.

www.ingramcontent.com/pod-product-compliance
Lightning Source LLC
Chambersburg PA
CBHW021932170526
45157CB00005B/2288